美猴王孙悟空的诞生

原著小说「西遊記」作者	吴承恩
故事改编	Kit Cheung, Ph.D. (Cantab)
美术设计	Glenn Burnett
游戏设计	Lorna Ayton, Ph.D. (Cantab)
游戏审核	David Whitebread, Ph.D.
翻译·编辑排版	Ko-Han Lai
中文校对	Edith Liu
英文校对	Marie Forsyth, MA (Oxon)

1

距今几千年前，有一座山，名为「花果山」。

Thousands of years ago, there was a mountain called "Flower and Fruit Mountain".

2

山顶上躺着一块石头，石头上有9个大洞和8个小洞。

On top of the mountain lay a stone. There were 9 big holes and 8 tiny holes in the stone.

Q1

石头坐在繁星点点的天空下 – A）天空中有多少颗星星？B）每颗星星有几个角？C）总共有多少个角？

The stone is sitting under the starry sky – *A) How many stars are there in the sky? B) How many corners does each star have? C) How many corners are there in total?*

A1

答案：
A)_____ B)_____ C)_____

Answers:
A)_____ B)_____ C)_____

3

石头已经在那里很多很多年。它每天吸收着日月精华,逐渐变得有灵气。

The stone lay there for many years and collected all the goodness from heaven, and it became magical.

4

有一天,这块神奇的石头怀孕了!它生了一颗石蛋。然后,一阵大风吹在石蛋上。

One day this magic stone was expecting a baby! It laid a stone egg. Later, a big wind blew on the egg.

5

忽然之间,石蛋变成了一只石猴子。他有两条胳臂和两条腿,一张脸,两只耳朵,两只眼睛,一个鼻子和一张嘴,看上去就像一只普通的猴子!

Suddenly it turned into a stone monkey, with 2 arms and 2 legs, a face, two ears, and two eyes, a nose and a mouth, like an ordinary monkey!

6

石猴子很快就学会爬行和走路!他走到土地的四个角落,向東南西北四个方向鞠躬行礼。

The monkey quickly learnt to crawl and walk. He walked first to the corners of the land in four directions, east, south, west, north. He bowed to each direction in turn.

Q2

石猴子依照东、南、西、北的顺序向四个角落鞠躬行礼。*A)你能说出两个有四个角的形状吗？B)你能画出你说的两个形状吗？*

The stone monkey bows to the four corners in the direction of east, south, west, north. *A)Can you name two shapes that have four corners? B)Can you draw the two shapes you named?*

A2

答案：

A)_____ B)_____

Answers:

A)_____ B)_____

7

他凝视天空，两道耀眼的金光从他的眼睛射入了天上的皇国。

The monkey gazed at the sky. Two bright golden beams of light shot out of his eyes into the kingdom in heaven.

8

天上的玉皇大帝对那两道耀眼的光束，惊讶不已，随即召唤了所有天神到凌霄宝殿。

The Jade Emperor in heaven was astonished by the two beams. He summoned all the gods to his palace.

9

他派出了千里眼天神（这位天神可以看得到数千里之外的事物）和顺风耳天神（这位天神可以听到千里之外的声音），察看人间发生了什么事。

He sent out the god with "Thousand Mile Eyes" (who could see things thousands of miles away) and the god with "Thousand Mile Ears" (who could hear whispers thousands of miles away) to see what had happened on the earth.

10

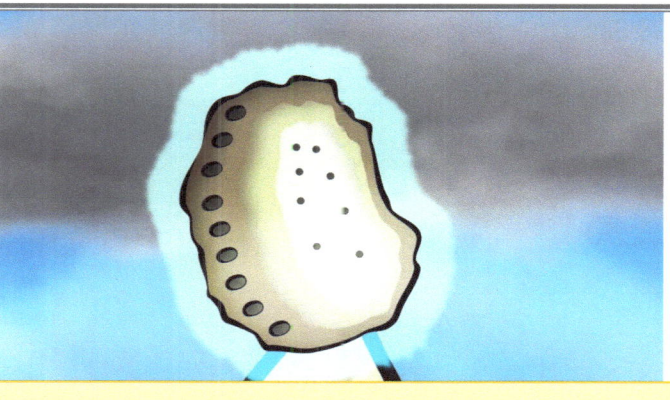

这两位天神报告说，在花果山上有一块神奇的石头生了一只石猴子。他们说：

These two gods reported that there was a magic stone on the Flower Fruit Mountain which had given birth to a stone monkey. They said,

11

「石猴子凝望天宫时，那两道耀眼的光束就从他眼睛里射上来。他现正在喝水和吃水果。光束很快便会消失了。」

"The bright beams came from the stone monkey's eyes when he gazed into heaven. He is now drinking water and eating fruits. The beams will soon disappear."

12

玉皇大帝松了一口气，没有再进一步勘查。

The Jade Emperor was relieved and did not pursue the matter further.

13

那只石猴子学会了如何走路，跑步，蹦跳和弹跃。他并在山中与不同种类的野生动物交朋友。

The stone monkey learned how to walk, run, jump and skip. He also made friends with different kinds of wild animals in the mountain.

14

晚上，他睡在石崖的底部。

At night, the stone monkey slept at the bottom of a stone cliff.

Q3

丛林里有很多动物啊！A）你能数出老虎，狼和猴子的数量吗？B）总共有多少只动物？

There are so many animals in the jungle! *A)Can you count the number of tigers, wolves, and monkeys? B) How many animals are there in total?*

A3

答案：
A)_____, _____, _____ B)_____

Answers:
A)_____, _____, _____ B)_____

Q4

帮助石猴子上山 – 每条路径都标有数字，石猴子只能使用2的倍数的路径。*他应该走哪条路径呢？*

Help the stone monkey to get up the mountain – each path is labelled with a number, the stone monkey can only use paths that are multiples of 2. *Which path should he take?*

A4

答案:
_____

Answers:
_____

15

一天，所有猴子们在一棵大树下嬉戏。有些猴子跳来跳去，有些采摘橘子，有些在追逐蜻蜓。他们边玩边笑，在溪边喝水，十分开心。

One day, all the monkeys played near a big bush. Some monkeys jumped around, some picked oranges, some chased dragonflies, they all played, laughed, and drank water from the stream and were very happy.

Q5

将树上的橘子平分给三只猴子：
A）每只猴子会得到多少颗橘子？B）如果有另一只猴子加入，那么每只猴子会得到多少颗橘子？

Share the oranges on the tree equally among 3 monkeys. *A) How many oranges does each monkey get? B) If another monkey came along to join in, how many oranges would they each have then?*

答案：
A) ＿＿＿只猴子，各得到＿＿＿颗橘子。
B) ＿＿＿只猴子，各得到＿＿＿颗橘子。

Answers:
A) ＿＿＿oranges for ＿＿＿monkeys.
B) ＿＿＿oranges for ＿＿＿monkeys.

然后他们都在溪边洗澡。一只猴子问道：「这条溪不知是从哪里流过来的？就让我们去上游寻找溪水的源头吧！」

Then the monkeys all took a bath in the stream. One monkey asked, "Do you know where this stream comes from? Let us go to the upper stream to find out where the source of the stream is!"

他们沿着溪流，走到了一个巨大的瀑布前，后方有一个洞穴，使瀑布看起来像垂挂在洞穴前的窗帘。

They all followed the stream and came to a great waterfall. Behind the waterfall was a cave. The waterfall appeared to form a curtain in front of the cave.

这些猴子们大声叫嚷着：「这溪流连贯了山脚至大海啊！」

The monkeys were shouting, "The water connects with the foot of the mountain and with the sea!"

19

然后，一只猴子说：「任何有能力进入大瀑布后面的洞穴，找到它的源头而没有受伤的猴子，我们就推举他成为我们的大王！」。

Then one monkey said, "If anyone can get into the cave and find the source of the water without getting hurt, we will honor him as our King!"

20

「好！好！好！」，所有的猴子们都附和着。突然间，石猴子大喊：「我去！我去！」

"Yes, Yes, Yes!" shouted all the monkeys. Then suddenly the stone monkey called out: "I will go, I will go!"

21

石猴子蹲下，向前猛力一跃，跳进了瀑布。

The stone monkey squatted, and made a great jump forward. He landed inside the waterfall.

22

洞穴里没有水。他眼前尽是石材家具，有石椅、石桌和石盘，就像一個家！

There was no water in the cave. He saw a living space with stone furniture. There were stone chairs, stone tables, and stone plates, everything needed for a home!

Q6

A）你看到多少把石椅？B）你看到多少张石桌？C）总共有几件家具？写下显示石材家具总数的算式。

A)How many stone chairs do you see? B)How many stone tables do you see? C)How many items of furniture are there in total? Write down the equation that shows the total number of stone furniture items.

A6

答案:

A)____ B)____ C)____ 总和:_____= ____

Answers:

A)____ B)____ C)____ Total:_____= ____

23

这里看起来像曾经有人住过。一切都很干净。

This living space looked as if somebody had lived there before. Everything was very clean.

24

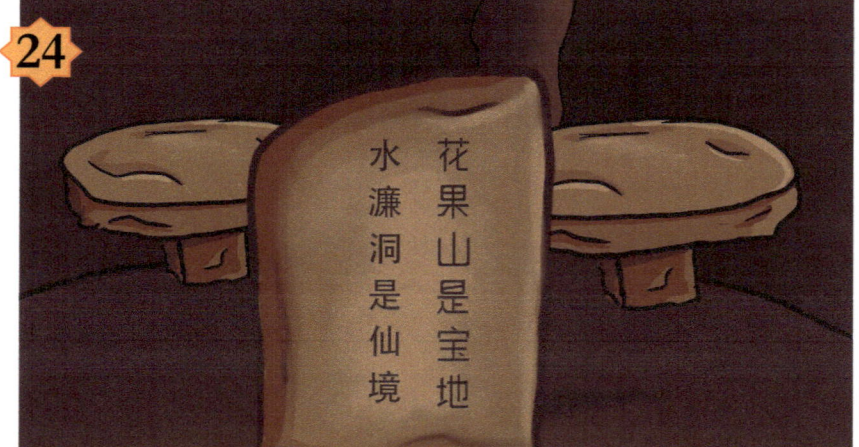

在房间的正中央，有块石头，上面写着：「花果山是宝地，水濂洞是仙境。」

In the middle of the room, there was a stone on which was written "Flower and Fruit Mountain is a fortunate land, Water Curtain Cave is a wonderful place."

25

石猴子很高兴,他立刻回到瀑布的入口处,然后跳回去会合其他猴子。

The stone monkey was delighted. He immediately went back to the waterfall, and then jumped back out to meet the monkeys.

26

猴子们看到石猴子毫无损伤地出现,非常兴奋。石猴子高兴地说:「里面没有水,这是一个可以为我们挡风遮雨的好住所。」

When the monkeys saw the stone monkey coming out unharmed, they were very excited. The stone monkey happily said, "there is no water inside and it is a good shelter for us to live in!"

27

猴子们更兴奋了,跟着石猴子进入瀑布后面的洞穴。

The monkeys became more excited and followed the stone monkey to enter the cave behind the waterfall.

28

他们进去后,都迫不及待地去触摸石家具,和躺在石床上。他们到处把玩所有的石头物品,直到全都玩累了。

Once they were inside, they all rushed to touch the stone furniture and to lie on the stone beds. They played around with all the stone items until they were exhausted.

Q7

所有的猴子都安全地在山洞里！看，他们都在玩！*A）你看见多少只猴子？* 每只猴子都有两只手和两只脚，玩他们的新玩具，但是有些顽皮的猴子却把他们的手脚藏了起来！*B）总共有多少只手和脚被藏了起来？*

All the monkeys are safe in the cave! Look at them all playing. *A) How many monkeys can you see?* Each monkey has 2 hands and 2 feet to play with their new toys, but some cheeky monkeys are hiding their hands and feet! *B) How many hands and feet are hidden in total?*

A7

答案：
A)_____ B)_____

Answers:
A)_____ B)_____

29

之后，石猴子对他们说：「你们之前承诺，无论谁能进入瀑布而毫无伤，你们就会以他为王，那你们现在应该尊称我为大王。」

Then the stone monkey spoke to them. "You have promised that whoever got into the waterfall without getting hurt would be made your king. Now you should call me King."

30

所有的猴子们都很高兴地向石猴子参拜，喊着：「大王！」

All the monkeys happily bowed to the stone monkey, calling him King.

11

石猴子也很开心,他删掉了「石」字,改称自己为「美猴王」。

The stone monkey was also happy, and he took away the word "stone" and named himself the "Handsome Monkey King".

时间过了 300 年,然后 500 年,美猴王和他的猴子们非常快乐地生活在「水濂洞」中,整天吃喝玩乐。

Time passed for 300 to 500 years, while Monkey King and his monkeys lived happily in the "water curtain cave", playing and eating all the time.

有一天,美猴王说:「终有一日,我会老死,所以我是无法永远幸福地生活在这里的。」所有的猴子们都难过地哭了起来。

One day, Monkey King said: "One day, I will get very old and die, so I cannot live here happily forever." The monkeys became sad and cried.

有一只老猴子建议美猴王去找高人拜师,指点迷津,学习如何成为圣人或神仙,便可以永远长生不老。

One old monkey suggested that Monkey King could go and find a mentor to learn how to become a saint or a fairy, because they can live forever.

35

美猴王决定要前往能找到师父的地方，学会长生不老之术。

Monkey King decided that he would go off to find a mentor to learn how to live forever.

36

第二天，猴子们帮美猴王建造了一只木筏。

On the following day the monkeys helped Monkey King to make a raft.

Q8

他们在木筏上放了些水果。*A）有多少颗橘子？B）有几根香蕉？C）总共有多少水果？*写下显示木筏上水果总数的算式。

They put some fruit onto the raft. *A) How many oranges are there? B) How many bananas are there? C) How many items of fruit are there in total?* Write down the equation that shows the total number of fruit on the raft.

A8

答案：

A)____ B)____ C)_____ 总和：_____ = _____

Answers:

A)____ B)_____ C)_____ Total: _____ = _____

13

然后,美猴王告别了他的猴子们,漂入了大海。

Then Monkey King said goodbye to his monkeys and drifted away into the big sea.

非"结局"!

欲知后事如何,且听下回分解!

Not "The End"!

If you want to know what will happen to Monkey King, please keep an eye out for our next episode!

答案

A1: A) 5 B) 10 C) 50
A2: A) 正方形/矩形/平行四边形/菱形/梯形/风筝形
B) ☐ ▭ ◇ ▽ △
A3: A) 2 只老虎、3 只狼、3 只猴子 B) 8
A4: 2-8-4-6
A5: A) 3 只猴子分得 4 颗橘子
B) 4 只猴子分得 3 颗橘子
A6: A) 2 B) 2 C) 4,总和: 2 + 2 = 4
A7: A) 6 B) 8
A8: A) 4 B) 6 C) 10 总和: 4 + 6 = 10

Answer Key

A1: A) 5 B) 10 C) 50
A2: A) Square/Rectangle/Parallelogram/Rhombus/Trapezoid/Kite
B) ☐ ▭ ◇ ▽ △
A3: A) 2 Tigers, 3 Wolves, 3 Monkeys B) 8
A4: 2-8-4-6
A5: A) 4 oranges for 3 monkeys B) 3 oranges for 4 monkeys
A6: A) 2 B) 2 C) 4, Total: 2 + 2 = 4
A7: A) 6 B) 8
A8: A) 4 B) 6 C) 10, Total: 4 + 6 = 10

家長支援小筆記

Q1 这个游戏有助于解答以下问题：- 数数，加法和乘法。

说说看: 游戏如何引入数学问题。

解答下一个问题前，读者首先需要解答基本问题（数数和加法）。有五颗星星，那么，每颗星星有 10 个角（包括一颗星星里的外角和内角）。此重复加法与乘法相同。讨论对较大的数字而言，乘法如何比加法更快。

更多帮助方式: 讨论物品的集合，例如书本或家里的盒子，并询问孩子总共有多少物品。例如，你有几本书？你有几个盒子？然后，问您的孩子这些书总共有几个角，或者盒子总共有几条边。

Q2 这个游戏有助于解答以下问题：- 形状识别和形状特征识别。

说说看: 不同的形状及其相似和不同的特征。例如，正方形，矩形，平行四边形，钻石形，梯形和风筝形，都具有四条边和四个外角，但是它们的形状不同。它们每个角的边长和角度构成了差异。

更多帮助方式: 鼓励在家中比较不同形状。大门，烤箱，烤面包机，冰箱，浴缸，和马桶等的形状是什么？

Notes for Parents

Q1 This helps to solve problems in: - counting, addition, and multiplication.

Talk about: how the game introduces the math problem. The reader first needs to solve the basic problem (counting and addition) before solving the next problem. There are five stars, then, there are 10 corners in each star (including both the outer and the inner corners within a star). This repeated addition is the same as multiplication. Discuss how for bigger numbers it is quicker to multiply than to add.

More ways to help: Talk about collections of objects, such as books, or boxes around the home, and ask your child how many objects there are in total. For instance, how many books do you have? How many boxes do you have? Then, ask your child how many corners do the books have altogether, or how many sides do the boxes have altogether.

Q2 This helps to solve problems in: - shapes recognition and identification of shape features.

Talk about: different shapes and their similar and different features. For instance, square, rectangle, parallelogram, rhombus, trapezoid and kite all have four sides and four outer corners but their shapes are different. Their length of each of their sides and angles at each corner constitute the differences.

More ways to help: Encourage comparing different shapes at home. What is the shape of the main door, the oven, the toaster, the refrigerator, the bath tub, the toilet bowl…etc?

家長支援小筆記

Q3 这个游戏有助于解答以下问题：-数数，加法和简单乘法。

说说看： 请参考前面的问题 – 使用加法计算总和。算出动物的数目后，孩子可以加总。还可以要求孩子以其他方式计算（通过使用乘法，更进阶），首先找到每种动物的总数，然后将每种动物的总数相加。

更多帮助方式： 讨论家中物品的集合。例如，计算衣橱里的T恤衫和袜子，或计算在厨房里的汤匙，筷子，碗，盘子。也询问孩子物品的总数。视觉物品能帮助数数。

Q4 这个游戏有助于解答以下问题：-2的倍数。

说说看： 识别数字并查看彼此之间是否存在任何关系。

更多帮助方式： 作为2的倍数的意义为何（即在2倍表中）。询问孩子是否能背诵2倍表，然后问："这些是2倍表中的数字吗？"

更多帮助方式： 鼓励孩子以2倍数数。让孩子在三个人之间分享诸如葡萄或葡萄干之类的物品 – 鼓励他们在分发物品时以两个物品为一组进行分享。

Notes for Parents

Q3 This helps to solve problems in: - counting, addition and simple multiplication.

Talk about: Refer to the earlier question – to find a total you use addition, once you have worked out how many animals you have and the child can add them together. You can also ask your child to do it in an alternative way (which is more advanced with multiplication) by first finding the total number of animals for each kind of animals, then, add the total number of each kind of animals together.

More ways to help: Talk about collections of objects around the home. For example, count tee-shirts and socks in the closet, or spoons, chopsticks, bowls, plates, in the kitchen. Also, ask your child how many objects there are in total. Visual objects can be used to help with counting.

Q4 This helps to solve problems in: - a multiple of 2.

Talk about: Identifying numbers and to see if there is any relationship between one and another.

More ways to help: What being a multiple of 2 means (i.e. in the 2 times table). Ask if the child can recite the 2 times table, then ask: 'Are these the numbers in the two times table?'

More ways to help: Encourage your child to count in two. Ask the child to share items, such as grapes, or raisins, between three people – encourage them to do it using two as a group when they give out the items.

家長支援小筆記

Q5 这个游戏有助于解答以下问题：-乘法和除法。

说说看： 被一个数字除意味着什么，这与在该数目的猴子之间分享果实是相同的。讨论除法和乘法（重复相加相同数字）如何相反。

更多帮助方式： 使用物品向孩子展示与乘法和除法之间的关系。拿 12 个物品，每 4 个排一排，对齐排 3 行，形成一个矩形。要在 4 个人之间分享 12 个物品时，可以按列分（分 4 列，每列 3 个物品），而在 3 个人之间共享 12 个物品时，则可以按行分（分 3 行，每行包含 4 个物品）。使用其他数量的物品操作，并与孩子讨论所有数字之间的关系。

Q6 这个游戏有助于解答以下问题：-数数，加法和简单乘法。

说说看： 请参阅前面的问题 – 使用加法计算总和。算出家具的数目后，孩子可以加总。还可以要求孩子以其他方式计算。首先找到每种家具的总数，然后将每种家具的总数相加。

更多帮助方式： 讨论家中物品的集合。例如，计算衣橱里的 T 恤衫和袜子，或计算在厨房里的汤匙，筷子，碗，盘子。也询问孩子物品的总数。视觉物品能帮助数数。

Notes for Parents

Q5 This helps to solve problems in: – multiplication and division.

Talk about: what it means to divide by a number, and that is the same as sharing the fruits between that number of monkeys. Discuss how division and multiplication (repeated adding of the same number) are the opposite of each other.

More ways to help: Show your child the relationship between multiplication and division with objects. Take 12 items and form a rectangle with them by lining up 3 rows of 4 objects. To share the 12 objects between 4 people you would separate by columns (4 columns with 3 objects in each), and to share 12 objects between 3 you would separate by rows (3 rows with 4 objects in each). Do this with other numbers of objects, talk to your child about the relationship between all the numbers.

Q6 This helps to solve problems in: - counting, addition and simple multiplication.

Talk about: Refer to the earlier question – to find a total, you use addition. Once you have worked out how many furnitures you have, then the child can add them together. You can also ask your child to do it in an alternative way by first finding the total number for each kind of furnitures, then, add the total number of each kind of furnitures together.

More ways to help: Talk about collections of objects around the home. For example, count tee-shirts and socks in the closet, or spoons, chopsticks, bowls, plates, in the kitchen. Also, ask your child how many objects there are in total. Visual objects can be used to help with counting.

家長支援小筆記

Q7 这个游戏有助于解答以下问题：-加法，减法，乘法和逻辑推理。
说说看： 首先，询问每只猴子各缺多少只手和脚，然后将这些数字相加，得出缺失的总数。想要更进阶的话，询问孩子根据图片中猴子的数量，应有多少手和脚（乘法），然后减去实际看到的数目。
更多帮助方式： 询问孩子"隐藏数字"-这些是可能无法具体计算的东西，例如已经吃过的糖果！例如，如果一开始有 15 颗糖果，每个人都吃了 3 颗，总共吃了多少颗？使用食品是一种很好的方法，意味着孩子不能再简单地计算出食品的数量，因为它们已经不存在了！

Q8 这个游戏有助于解答以下问题：-数数，加法和简单乘法。
说说看： 请参阅前面的问题 – 使用加法计算总和。算出水果的数目后，孩子可以加总。还可以要求孩子以其他方式计算。首先找到每种水果的总和，然后将每种水果的总数相加。
更多帮助方式： 讨论家中物品的集合。例如，计算衣橱里的 T 恤衫和袜子，或计算在厨房里的汤匙，筷子，碗，盘子。也询问孩子物品的总数。视觉物品能帮助数数。

Notes for Parents

Q7 This problem involves:- addition, subtraction, multiplication, and logical reasoning.
Talk about: First ask how many hands and feet are missing for each individual monkey, then add these numbers together for the total number missing. To make it more advanced, ask your child how many hands and feet there should be in total based on the number of monkeys in the picture (multiplication). Then subtract how many can actually be seen.
More ways to help: Ask your child about 'hidden numbers' - these can be things that can't specifically be counted, like sweets that have already been eaten! For example if you started with 15 sweets, and you each ate 3, how many were eaten in total? Using food items is a good way to mean your child cannot simply count the items because they aren't there any more!

Q8 This helps to solve problems in: - counting, addition and simple multiplication.
Talk about: Refer to the earlier question – to find a total, you use addition. Once you have worked out how many fruits you have, then the child can add them together. You can also ask your child to do it in an alternative way by first finding the total number for each kind of fruits, then, add the total number of each kind of fruits together.
More ways to help: Talk about collections of objects around the home. For example, count tee-shirts and socks in the closet, or spoons, chopsticks, bowls, plates, in the kitchen. Also, ask your child how many objects there are in total. Visual objects can be used to help with counting.

游戏设计

萝尔娜·艾顿博士
剑数说™ 数学顾问长

艾顿博士花一半时间在剑桥的一所私立高中教 A-Level 学生，另一半时间在剑桥大学应用数学和理论物理系，使用数学模型开发新科技以降低飞机噪音。艾顿博士于 2007 年考入剑桥大学丘吉尔学院，攻读数学学士学位，接着攻读第三部分数学，然后继续攻读博士学位，并于 2014 年完成数学博士课程，取得数学博士学位。艾顿博士确保数学在剑数说™的游戏中得到广泛的应用，以及确保学习数学是有趣和有益的！艾顿博士为剑数说™设计和审核游戏。

Game Designer Bio

Lorna Ayton, Ph.D.
CAMathories™ Mathematics Advisor

Dr. Ayton spends half of her time teaching A-Level students at a private high school in Cambridge and the other half of her time in the University of Cambridge Department of Applied Mathematics and Theoretical Physics, where she uses mathematical modelling to develop new technologies to reduce aircraft noise. Dr. Ayton was admitted to Churchill College, University of Cambridge, in 2007 to study B.A. Mathematics, followed by Part III Mathematics, and then for a Ph.D. in Mathematics which she completed in 2014. Dr. Ayton ensures that fundamental mathematics concepts are well applied in the games and learning mathematics can be fun and rewarding. Dr. Ayton designs and reviews games for CAMathories™.

游戏审核

大卫·怀特伯瑞德博士
剑数说™ 教育顾问总监

怀特伯瑞德博士曾担任剑桥大学教育学院之「游玩之于教育，发展和学习研究中心」（PEDAL）的代理主任（对外关系）。怀特伯瑞德博士在儿童透过游戏学习的研究方面拥有超过30年经验。他是剑桥大学教育学院的创院成员之一。怀特伯瑞德博士为剑数说™设计和审核游戏。

Game Reviewer Bio

David Whitebread, Ph.D.
CAMathories™ Chief Education Advisor

Dr. Whitebread, had been the Acting Director (External Relations), Centre for Research on Play in Education, Development and Learning (PEDAL), in the Faculty of Education, University of Cambridge. Dr. Whitebread has more than 30 years of research experience in children learning through play. He is one of the founding members of the Faculty of Education at the University of Cambridge. Dr. Whitebread designs and reviews games for CAMathories™.

剑数说™《孙悟空的诞生》改编自《西游记》，是西元 1592 年中国明朝匿名出版的中国四大古典小说之一。有些人认为曾生活在那个时期的吴承恩，是小说的作者。

虽然作者的身份仍然不确定，但"美猴王"的故事已成为东亚最受喜爱的故事之一。小说描述了唐代和尚，唐三藏，的一次朝圣冒险经历。他从中国长途跋涉至中亚和印度，以获取神圣的佛陀经文。本书的主角"美猴王"孙悟空在协助唐三藏完成这一传奇而刺激的任务中，扮演了重要的角色。

小说的每一章节均富含想像力，充满着各种神灵、女神、动物与人类之间的惊奇互动。本书叙述了孙悟空是如何诞生，并成为"美猴王"。

本书旨在通过与美猴王和他的猴群一起玩游戏，帮助 5 至 8 岁的孩子学习数学。石猴是怎么样出生的？他是如何爬上山丘的？猴子们总共吃了多少水果？石猴如何成为美猴王？美猴王和他的猴子们如何进出水帘洞？

本书中的游戏引导年幼的孩子们帮助猴子们获得足够的水果吃，找到通往山顶的正确路径，到达山洞作为居所，并跳入他们的新家！猴子们需要分发正确数量的水果以平均分配，计算到达山顶和山洞的正确步伐数字，选择正确的数字和符号来完成序列，并建立方程式以进展到下一步。

CAMathories™ The Birth of Monkey King was adapted from "The Journey to the West", which is one of the four Great Chinese Classical Novels published anonymously during the Ming dynasty in China in 1592. Some people believed that WU Cheng En, who lived during that period, was the author of the novel.

The authorship remains uncertain but the story of Monkey King became one of the most popular stories in East Asia. The novel describes an adventurous pilgrimage made by a Tang dynasty Buddhist monk, TANG San Zang. He traveled from China to Central Asia and India to obtain a sacred Buddha text. The main character who appears in this book, SUN Wu Kong, the "Handsome Monkey King", played a key role in assisting TANG in this legendary and exciting quest.

Each chapter of the novel is highly imaginative and full of amazing interaction between mythical gods and goddesses, animals and human characters. The book shows how Monkey King was born and became the "Handsome Monkey King."

This book aims to help 5 to 8-year-old children to learn mathematics by playing games with the Handsome Monkey King and his monkeys. How was the stone monkey born? How did he climb up the hills and mountains? How many fruits did the monkeys eat in total? How did the stone monkey become the Handsome Monkey King? How did the Handsome Monkey King and his monkeys get in and out of the Water Curtain Cave?

The games in this book involve young children in helping the monkeys to get enough fruits to eat, to find the right path to the top of the mountain, and to the cave for shelter, and to jump into their new home! The monkeys need to split the right number of fruits to share equally, to count the right number of steps to reach the mountain top and the cave, to select the right numbers and symbols to complete a sequence, and to make an equation in order to get to the next step.

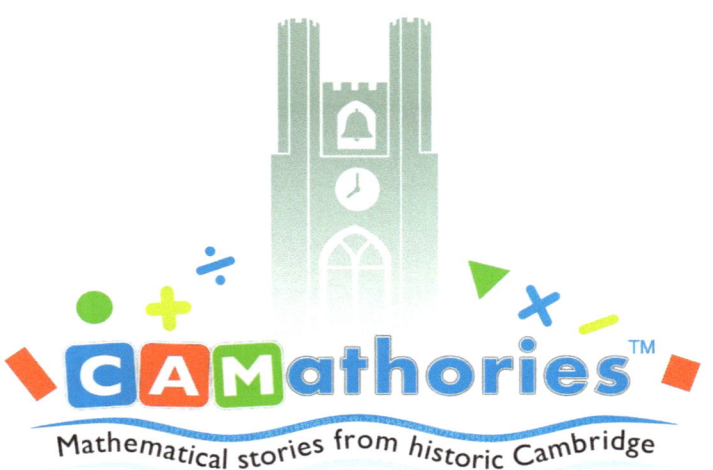

The Birth of Monkey King, Sun Wukong

Original Novel Author of the "Journey to the West"	Cheng En Wu
Story Adaptation	Kit Cheung, Ph.D. (Cantab)
Artist	Glenn Burnett
Game Designer	Lorna Ayton, Ph.D. (Cantab)
Game Reviewer	David Whitebread, Ph.D.
Editor & Translator	Ko-Han Lai
Chinese Proofreader	Edith Liu
English Proofreader	Marie Forsyth, MA (Oxon)

www.ingramcontent.com/pod-product-compliance
Lightning Source LLC
Chambersburg PA
CBHW041819080526

44587CB00004B/143